Straw

by Taneesha Campbell
illustrated by John Kurtz

HMH

Copyright © by Houghton Mifflin Harcourt Publishing Company

All rights reserved. No part of this work may be reproduced or transmitted in any form or by any means, electronic or mechanical, including photocopying or recording, or by any information storage and retrieval system, without the prior written permission of the copyright owner unless such copying is expressly permitted by federal copyright law. Requests for permission to make copies of any part of the work should be submitted through our Permissions website at https://customercare.hmhco.com/contactus/Permissions.html or mailed to Houghton Mifflin Harcourt Publishing Company, Attn: Intellectual Property Licensing, 9400 Southpark Center Loop, Orlando, Florida 32819-8647.

Printed in the U.S.A.

ISBN 978-1-328-77228-2

4 5 6 7 8 9 10 2562 25 24 23 22 21

4500844736   A B C D E F G

If you have received these materials as examination copies free of charge, Houghton Mifflin Harcourt Publishing Company retains title to the materials and they may not be resold. Resale of examination copies is strictly prohibited.

Possession of this publication in print format does not entitle users to convert this publication, or any portion of it, into electronic format.

Grandpa wanted strawberries.
He wanted lots of berries.
He asked Tina and Theo to get some.

Tina had 1 bucket full.

So did Theo.

Each bucket had 10 berries.

How many berries did they pick?

Tina kept on picking.
Theo kept on picking.
They filled 4 buckets.

Now how many berries did they pick?

Tina filled 2 more buckets.
Theo filled 2 more buckets.
They now had 4 buckets each.

How many berries did they pick?

They picked 2 more buckets.
Then they were done.
Tina and Theo started home.

How many berries are in the wagon?

They gave Grandpa the berries.
Tina and Theo gave him 5 buckets
each. Grandpa ate one berry.
"Mmm. These will make a
good pie."

How many berries did each fox give him?

Responding

Math Concepts

Buckets of Berries

Draw
Look at page 5. Draw the buckets you see in the wagon.

Tell About
Look at page 5. Tell how many buckets of strawberries the foxes picked. Tell how many berries they have in all.

Write
Draw Conclusions Look at page 5. Write how many strawberries the foxes picked in all.